Get to Know Big Cats

LIONS

By Bray Jacobson

Please visit our website, www.garethstevens.com. For a free color catalog of all our high-quality books, call toll free 1-800-542-2595 or fax 1-877-542-2596.

Library of Congress Cataloging-in-Publication Data
Names: Jacobson, Bray, author.
Title: Lions / Bray Jacobson.
Description: Buffalo, New York : Gareth Stevens Publishing, [2024] | Series: Get to know big cats | Includes index. | Audience: Grades K-1
Identifiers: LCCN 2022045106 (print) | LCCN 2022045107 (ebook) | ISBN 9781538286081 (library binding) | ISBN 9781538286074 (paperback) | ISBN 9781538286098 (ebook)
Subjects: LCSH: Lion–Juvenile literature.
Classification: LCC QL737.C23 J3373 2024 (print) | LCC QL737.C23 (ebook) | DDC 599.75/5–dc23/eng/20220923
LC record available at https://lccn.loc.gov/2022045106
LC ebook record available at https://lccn.loc.gov/2022045107

First Edition

Published in 2024 by
Gareth Stevens Publishing
2544 Clinton Street
Buffalo, NY 14224

Editor: Kristen Nelson
Designer: Leslie Taylor

Photo credits: Cover, p. 1 ArtMediaFactory/Shutterstock.com; p. 5 YOGESH BHATIA PHOTOARTIST/Shutterstock.com; p. 7 Henrico Muller/Shutterstock.com; pp. 9, 24 (mane) Peter Betts/Shutterstock.com; p. 11 AndyElliott/Shutterstock.com; p. 13 PHOTOCREO Michal Bednarek/Shutterstock.com; p 15 oNabby/Shutterstock.com; pp. 17, 24 (cub) Dave Pusey/Shutterstock.com; pp. 19, 24 (lioness) Tom Voosen/Shutterstock.com; p. 21 Mogens Trolle/Shutterstock.com; p. 23 Papa Bravo/Shutterstock.com.

Printed in the United States of America

CPSIA compliance information: Batch #CSGS24: For further information contact Gareth Stevens at 1-800-542-2595.

Contents

Lions are big cats!

They have fur.
It is yellow, red,
or brown.

Male lions have fur around their head. This is a mane.

All lions have claws.
There is one
on each toe.

Most lions live in Africa.

Lions live in groups.
These are prides.

Prides have mother lions.
Lion babies are cubs.

Lionesses hunt for food.

They hunt zebras.
They hunt wild pigs.

Lions sleep a lot!

Words to Know

claw

cub

mane

Index